FOOD SCIENCE AND TECHNOLOGY

SPRAY DRYING OF ACAI (EUTERPE OLERACEA MART) JUICE: EFFECT OF PROCESS VARIABLES AND TYPE OF CARRIER AGENT ON PRODUCTS QUALITY AND STABILITY

FOOD SCIENCE AND TECHNOLOGY

Additional books in this series can be found on Nova's website
under the Series tab.

Additional E-books in this series can be found on Nova's website
under the E-books tab.

FOOD SCIENCE AND TECHNOLOGY

SPRAY DRYING OF ACAI (EUTERPE OLERACEA MART) JUICE: EFFECT OF PROCESS VARIABLES AND TYPE OF CARRIER AGENT ON PRODUCTS QUALITY AND STABILITY

RENATA V. TONON
CATHERINE BRABET
AND
MÍRIAM D. HUBINGER

NOVINKA

Nova Science Publishers, Inc.
New York

For permission to use material from this book please contact us:
Telephone 631-231-7269; Fax 631-231-8175
Web Site: http://www.novapublishers.com

Additional color graphics may be available in the e-book version of this book.

LIBRARY OF CONGRESS CATALOGING-IN-PUBLICATION DATA

Tonon, Renata V.
Spray drying of agai (Euterpe oleracea Mart) juice : effect of process variables and type of carrier agent on product's quality and stability / authors: Renata V. Tonon, Catherine Brabet, Miriam D. Hubinger.
p. cm.
Includes index.
ISBN 978-1-61122-421-4 (softcover)
1. Fruit juices--Drying. 2. Spray drying. 3. Agam palm. I. Brabet, Catherine. II. Hubinger, Miriam D. III. Title.
TP562.T66 2011
664'.8042--dc22

2010041595

Published by Nova Science Publishers, Inc. † New York

CONTENTS

PREFACE

This book describes and discusses some results obtained through the study of the microencapsulation of açai juice by spray drying using different carrier agents. Initially, the influence of process conditions on the moisture content, process yield and anthocyanin retention was evaluated using a central composite design. From the conditions selected in this first section (inlet air temperature of 140°C, feed flow rate of 15 g/min and 6% of carrier agent), particles were produced using four types of carrier agents: maltodextrin 10DE, maltodextrin 20DE, gum Arabic and tapioca starch. These particles were then characterized with respect to water activity, bulk and absolute density, porosity, particle size distribution and morphology. The samples produced with maltodextrin 20DE and with gum Arabic exhibited the highest water activity and the lowest particles size, while those produced with tapioca starch were the less porous, with the lowest bulk density and highest mean diameter. Then, physical stability of particles, when exposed to different relative humidities, was evaluated through the construction of sorption isotherms and determination of the glass transition temperature. The samples produced with maltodextrin 10DE exhibited the highest critical water activity, being considered as the most stable powder. Glass transition temperature decreased with increasing moisture content, confirming the plasticizant effect of water on this property. Finally, anthocyanin stability of powders stored at different temperatures and relative humidities was evaluated. The increase of both these parameters resulted in higher anthocyanin degradation. Maltodextrin 10DE was the carrier agent that showed the best pigment protection, for all the conditions studied.

Chapter 1

INTRODUCTION

Açai (*Euterpe oleracea* Mart.) is a typical palm berry from Amazonia, with great occurrence and economical importance in the Brazilian state of Pará. It consists of a round fruit, with a proportionally big seed (approximately 85%), little pulp content and diameter varying between 1.0 and 1.5 cm (Figure 1). Besides having high energetic content, açai is also rich in fibers, vitamin E, proteins, minerals and unsaturated fatty acids such as Omega-6 and Omega-9. When processed into juice, açai has a large market in the Amazonian region, with a consumption of 100,000-180,000 liters per day, being used to produce energetic beverages, ice cream, jelly and liquor.

Figure 1. Açai fruit before harvest.

In the last years, açai has been recognized for its functional properties for use in food and nutraceutical products, due to its high antioxidant activity, which is related to its high anthocyanin and phenolic content (Coisson et al., 2005; Schauss et al., 2006). The presence of anthocyanins has increased the attractiveness of açai, not only for its functional properties, which are encouraging many industries to produce açai capsules, but also because this fruit can be considered to be an important source of natural pigments, which have no toxic effects and can contribute to the reduction of synthetic pigments in foods. This explains the significant growth of açai production in the state of Pará, which increased from 257,000 tons in 2003 to 581,000 tons in 2008 (SAGRI, 2010), being commercialized within various Brazilian states, as well as being exported to other countries around the world.

However, due to its high perishability, açai has a short shelf life, even under refrigeration. Moreover, anthocyanins are very unstable in processing and storage. Thus, the food industry is constantly looking for processes that increase its shelf life and improve pigment stability.

The production of açai powder represents an alternative, aiming at improving the product's preservation. Powdered fruit juices have low water activity, which makes difficult or even hinders microorganism growing and deterioration reactions, thus increasing its shelf-life. In addition, the production of powdered fruit juices also has advantages, like the ease of transport, storage and handling of the final product, even for direct consumption or as an ingredient in the elaboration of other food products.

The process generally used in the production of powdered fruit juices is spray drying (Quek et al., 2007; Cano-Chauca et al., 2005; Righetto & Netto, 2005; Dib Taxi et al., 2003). It consists of the atomization of the liquid in a chamber that receives a hot air flow in such a way that the fast water evaporation allows for maintaining a low temperature inside the particles. The physicochemical properties of powders produced by spray drying depend on some process variables, such as the characteristics of the liquid feed (viscosity, particles size, flow rate) and of the drying air (temperature, pressure), as well as the type of atomizer. Therefore, it is important to optimize the drying process in order to obtain products with better sensory and nutritional characteristics and better process yield.

Despite all the advantages related to the spray drying process, the powders resulting from drying of fruit juices usually exhibit problems like stickiness and high hygroscopicity, due to the presence of low molecular weight sugars and acids, which have low glass transition temperature. Some of these problems can be solved by the addition of carrier agents to the product before atomization, such as carbohydrates (maltodextrins, starch products, gums, and cellulose), proteins and some lipids. Such agents, besides increasing powder T_g, are very useful for microencapsulation purposes.

Microencapsulation has been used in order to protect sensitive foods or ingredients against adverse conditions, increase their stability and promote controlled release. In the case of açai juice, microencapsulation can represent a promising technique, which can protect anthocyanins against factors like oxygen, light and heat (increasing its shelf life), besides resulting in better handling properties, providing lower water adsorption and making the product less hygroscopic.

Taking into account the potential of powdered açai juice as a natural colorant and its high nutritional value, this book describes and discusses some results obtained in the study of spray drying of açai juice, as well as the effect of different carrier agents on powder physicochemical properties and anthocyanin stability upon storage.

Chapter 2

Microencapsulation by Spray Drying

2.1. General Aspects

Microencapsulation is defined as a technique by which liquid droplets, solid particles or gas compounds are entrapped into thin films of a food grade carrier agent (Gharsallaoui et al., 2007).

According to Ré (1998), the main reasons for using microencapsulation in food products are: to protect food against adverse environmental conditions (light, moisture, oxygen, UV radiation) and nutritional losses, to incorporate controlled release mechanisms to the formulations, to mask or preserve flavors and aromas and, finally, to make the product more attractive, by promoting a better flexibility and control in the development of more tasty and nutritive products, in order to satisfy consumers' expectation.

There are many methods used for microencapsulation of food ingredients, which are divided in: physical (spray drying, freeze drying, spray cooling, spray chilling, spray coating, extrusion, fluidized bed drying, co-crystallization), chemical (interfacial polymerization, *in situ* polymerization) and physicochemical (simple and complex coacervation, solvent evaporation, molecular inclusion). The choice of a method depends on the economic factors, the core sensitivity, the desired microcapsule size, the physicochemical properties of the core and the wall material, as well as the mechanism of release (Jackson & Lee, 1991).

The most common process used for microencapsulation in the food industry is spray drying (or atomization), which is a well established and straightforward technology that results in particles with good quality (Gouin, 2004). Equipment is readily available and production costs are lower than most other methods. This process can be useful for drying heat-sensitive foods, pharmaceutical and other substances, since it promotes rapid water evaporation from the droplets, keeping low the temperature inside the particles.

Atomization results from the application of an energy which acts on the liquid until its disruption and disintegration, creating a spray of droplets. This spray is put in contact with the hot air and an immediate drying occurs, resulting in the collect of powder. According to Goula et al. (2004), in the early stages of drying, the droplet has a free liquid surface and evaporation from this surface is fast. The water removal cause the solute to be more concentrated at the surface, which depends on the evaporation speed and the rate at which the liquid can be replenished from the interior of the droplet. Due to this increased concentration, solids form a crust around a hollow particle, whose thickness depends on the drying rate (high initial drying rates lead to larger particles with thin shells, whereas, low initial drying rates lead to smaller particles with thicker shells).

2.2. Spray Drying of Fruit Juices

Spray drying, when performed at optimized conditions, has proved to be an efficient method for obtaining several types of food products. Drying of sugar-rich products such as fruit juices has great economical potential, since it results in products with very much reduced volume and longer shelf life, besides making their transport and storage easier.

However, fruit juice powders obtained by spray drying have some drawbacks in their handling properties, such as stickiness, higroscopicity and solubility, making their packaging and utilization substantially difficult. According to Bhandari et al. (1997), the sticky behavior of sugar and acid-rich materials is attributed to low molecular weight sugars such as fructose, glucose and sucrose, and organic acids such as citric, malic and tartaric acid, which constitute more than 90% of the solids in fruit juices and purees. According to the authors, the fast water removal that occurs during atomization results in a completely amorphous product, or even in a product with some micro-crystalline regions that disperses in an amorphous mass. The sugars and acids present in fruit juices have low glass transition temperature and are very hygroscopic in their amorphous state and lose free flowing nature at high moisture content (Roos & Karel, 1991).

Bhandari et al. (1997) described the main stages of spray drying, as well as the changes occurring in the sugar-rich products during drying. According to the authors, the droplets are initially dispersed individually in a large volume of the dryer, avoiding agglomeration of particles. Towards the lower part of the dryer, when the particles are already solid, agglomeration should neither occur. However, due to the presence of high sugar content with low T_g, the product can remain as syrup, even at low moisture content level. Depending on the product composition and the spray drying conditions, the surface of droplets may remain plastic, resulting in the stickiness to the drying chamber wall or even between the particles. Thus, the amorphous product obtained at the end of the drying process could be either syrup or a sticky powder, or even a relatively free-flowing powder (Bhandari & Howes, 1999).

In general, the glass transition temperature of sugar-rich foods is so low, that drying of these pure products is not economically viable. In this context, the addition of carrier agents with high molecular weight to the product before atomization is necessary, in order to increase its glass transition temperature and reduce powder stickiness.

2.3. Carrier Agents

The choice of the carrier agent depends on the physicochemical properties of the material to be dried, the process used to form the particle and its desired final properties. The ideal wall material must be non-reactive with the core material, exhibit low viscosity at high concentrations, be able to disperse or emulsify the active material and to stabilize the emulsion formed, entrap and keep the core material inside its structure during processing and storage. Moreover, it must completely release the core material (release can be controlled or not), provide the maximum protection against environmental conditions, be soluble in solvents acceptable in the food industry, show high availability and low cost (Gharsallaoui et al., 2007; Desai & Park, 2005).

Typical carrier agents include maltodextrins, hydrophobically modified starches, gum Arabic, milk and soy proteins and mixtures of these materials. Several works can be found in the literature showing maltodextrin and gum Arabic as carrier agents used in the spray drying of fruit juices, such as acerola (Righetto & Netto, 2005), mango (Cano-Chauca *et al.*, 2005), cactus pear (Rodríguez-Hernández et al., 2005) and camu-camu (Dib Taxi et al., 2003).

Maltodextrins are products of starch hydrolysis, consisting of α-D-glucose units linked mainly by (1→4) glycosidic bonds and described by their dextrose equivalency (DE), which varies with the hydrolysis degree, determines their reducing capacity and is inversely related to their average molecular weight (BeMiller & Whistler, 1996). They are very useful for spray drying of food materials, mainly due to their low cost. According to Reiniccius (2001), maltodextrins are mainly used in materials that are difficult to dry, like fruit juices, flavorings, enzymes or sweeteners.

Gum Arabic is a natural plant exudate of Acacia trees, comprising a complex hetero-polysaccharide with a highly ramified structure, with a main chain comprising D-galactopyranose units joined by β-D-(1→3) glycosidic bonds. Side chains with different chemical structures are linked to the main chain by β-(1→6) bonds (BeMiller & Whistler, 1996). Gum Arabic is the only gum used in food products that exhibits high solubility and low viscosity in aqueous solution, which facilitates the spray drying process.

Tapioca starch is a fine flour refined from cassava root (*Manihot esculenta* Crantz) by natural fermentation, which has been used by some Brazilian companies as a carrier agent in the production of powdered fruit juices, in order to ensure that no genetically modified material is being used in the powder production – a condition required by some importing countries. Moreover, the use of such an agent has other advantages, such as absence of flavor and taste, high availability and low cost in Brazil.

Chapter 3

SPRAY DRYING OF AÇAI JUICE

3.1. EFFECT OF PROCESS VARIABLES

In our study about microencapsulation of açai juice, we firstly evaluated the influence of some spray drying conditions (inlet air temperature, feed flow rate and maltodextrin concentration) on the powder physicochemical properties, according to a central composite design. For this, the açai pulp was filtered through a filter paper, in order to eliminate solids in suspension, facilitating the product's passage through the nozzle atomizer. Filtration increased the anthocyanin content per gram of dried mass and reduced the fat content (Table 1), making the product less susceptible to lipid oxidation. Then, maltodextrin 10DE was added to the filtered pulp under magnetic agitation, until complete dissolution.

Table 1. Composition of pure and filtered açai pulp (*Euterpe oleraceae* Mart.)

Analyzed item	Pure pulp	Filtered pulp	Analysis method
Moisture (% wet basis)	85.96 ± 0.11	96.94 ± 0.01	A.O.A.C. (1990)
Proteins (%)	1.43 ± 0.04	0.50 ± 0.05	A.O.A.C. (1990)
Lipids (%)	6.53 ± 0.03	0.21 ± 0.01	Bligh and Dyer (1959)
Fibers (%)	4.52 ± 0.22	n.d.*	A.O.A.C. (1990)
Total sugars (%)	0.48 ± 0.05	1.69 ± 0.12	A.O.A.C. (1990)
Ash (%)	0.44 ± 0.01	0.39 ± 0.01	A.O.A.C. (1990)
Acidity (% citric acid)	0.34 ± 0.02	0.32 ± 0.02	A.O.A.C. (1990)
Anthocyanins (mg/100g dried matter)	1265.14 ± 16.27	3920.48 ± 33.74	Francis (1982)

*n.d. = not detected.

Spray drying process was performed in a laboratory scale spray dryer LabPlant SD-05 (Huddersfield, England), with a 1.5 mm diameter nozzle and main spray chamber of 500 mm × 215 mm. The mixture was fed into the main chamber through a peristaltic pump and the feed flow rate was controlled by the pump rotation speed. Drying air flow rate was 73 m^3/h and compressor air pressure was 0.06 MPa.

Table 2 shows the codified independent variables of the central composite design.

Table 2. Codified independent variables of the central composite design

Variables	- 1.68	- 1	0	+ 1	+ 1.68
Inlet air temperature (°C)	138	150	170	190	202
Feed flow rate (g/min)	5	9	15	21	25
Maltodextrin concentration (%)	10	14	20	26	30

Moisture content, process yield and anthocyanin retention were analyzed as responses. Process yield was calculated as the relationship between total solids content in the resulting powder and total solids content in the feed mixture. Moisture contents were determined gravimetrically by drying in a vacuum oven at 70°C until constant weight (A.O.A.C., 1990) Total anthocyanin content was determined by the spectrophotometric method, according to Francis (1982). The experimental results obtained in the central composite design are shown in Table 3.

Data were fitted to the following polynomial:

$$y = \beta_0 + \beta_1 x_1 + \beta_2 x_2 + \beta_3 x_3 + \beta_{11} x_1^2 + \beta_{22} x_2^2 + \beta_{33} x_3^2 + \beta_{12} x_1 x_2 + \beta_{13} x_1 x_3 + \beta_{23} x_2 x_3 \quad (1)$$

where β_n are constant regression coefficients; y is the response (moisture content, process yield and anthocyanin retention), and x_1, x_2 and x_3 are the coded independent variables (inlet air temperature, feed flow rate and maltodextrin concentration, respectively).

Table 3. Powders moisture content, process yield and anthocyanin retention obtained for the 17 trials of the experimental design

Tests	Inlet air temperature (°C)	Feed flow rate (g/min)	Maltodextrin concentration (%)	Moisture content (%)	Process yield (%)	Anthocyanin retention (%)
1	150 (-1)	9 (-1)	14 (-1)	1.23±0.07	50.33	84.86±1.74
2	190 (+1)	9 (-1)	14 (-1)	0.69±0.08	52.47	81.40±2.04
3	150 (-1)	21 (+1)	14 (-1)	2.89±0.10	42.53	86.01±2.86
4	190 (+1)	21 (+1)	14 (-1)	0.86±0.04	49.32	79.09±2.78
5	150 (-1)	9 (-1)	26 (+1)	1.92±0.27	45.74	84.08±2.31
6	190 (+1)	9 (-1)	26 (+1)	0.64±0.02	48.17	79.38±1.05
7	150 (-1)	21 (+1)	26 (+1)	2.34±0.06	34.39	83.99±1.23
8	190 (+1)	21 (+1)	26 (+1)	1.05±0.07	38.38	79.29±1.66
9	138 (-1.68)	15 (0)	20 (0)	2.56±0.17	46.01	84.62±1.58
10	202 (+1.68)	15 (0)	20 (0)	0.66±0.18	55.66	77.21±1.10
11	170 (0)	5 (-1.68)	20 (0)	0.95±0.24	50.27	80.05±1.15
12	170 (0)	25 (+1.68)	20 (0)	1.88±0.02	39.02	80.82±1.37
13	170 (0)	15 (0)	10 (-1.68)	1.78±0.06	50.83	83.13±1.28
14	170 (0)	15 (0)	30 (+1.68)	1.68±0.40	41.40	84.06±3.15
15	170 (0)	15 (0)	20 (0)	1.29±0.05	46.04	84.65±1.44
16	170 (0)	15 (0)	20 (0)	1.45±0.10	48.85	82.42±2.33
17	170 (0)	15 (0)	20 (0)	1.30±0.11	47.21	79.76±1.43

Table 4 shows the regression coefficients for the coded second order polynomial equation, the *p* values and the determination coefficients (R^2). Some non-significant terms were eliminated and the resulting equations were tested for adequacy and fitness by the analysis of variance (ANOVA). The fitted models were suitable, showing significant regression, low residual values and no lack of fit.

Table 4. Coded second-order regression coefficients for process yield, moisture content and anthocyanin retention

Coefficients	Process yield (%)	Moisture content (%)	Anthocyanin retention (%)
β_0	48.49	1.48	82.05
β_1	2.31	–0.61	–2.36
β_2	–3.74	0.31	NS
β_3	–3.21	NS	NS
β_{11}	NS	NS	NS
β_{22}	–1.64	NS	NS
β_{33}	–1.12	NS	NS
β_{12}	NS	–0.19	NS
β_{13}	NS	NS	NS
β_{23}	–1.27	NS	NS
R^2	0.926	0.907	0.703
p-value	<0.0001	<0.0001	<0.0001

NS: Non-significant (p>0.05).

Figures 2 to 4 show the response surfaces generated from the models obtained by statistic analyses of the results of process yield, moisture content and anthocyanin retention.

Moisture content was significantly influenced by inlet air temperature and feed flow rate, varying from 0.64 to 2.89%, which is in agreement with the values reported by Papadakis et al. (2006) for spray-dried raisin juice. Inlet air temperature was the variable that showed the greatest influence on powders moisture content. Higher inlet air temperatures imply in higher temperature gradient between the atomized liquid and the drying air, resulting in a greater driving force for water evaporation and thus in products with lower moisture content. A similar behavior was observed by Quek et al. (2007), Rattes & Oliveira (2007) and Grabowski et al. (2006), studying the spray drying of watermelon juice, sodium diclofenac and sweet potato puree, respectively. The feed flow rate negatively affected powders moisture content. The higher the flow rate, the shorter is the contact time between the liquid and the drying air, which makes the heat transfer less efficient and results in lower water evaporation. Hong & Choi (2007) also observed a more pronounced effect of the drying temperature than of the pump rate on the moisture content of spray-dried protein-bound polysaccharide from *Agaricus blazei* Murill. As illustrated in Figure 2(b), maltodextrin concentration did not influence powders moisture content.

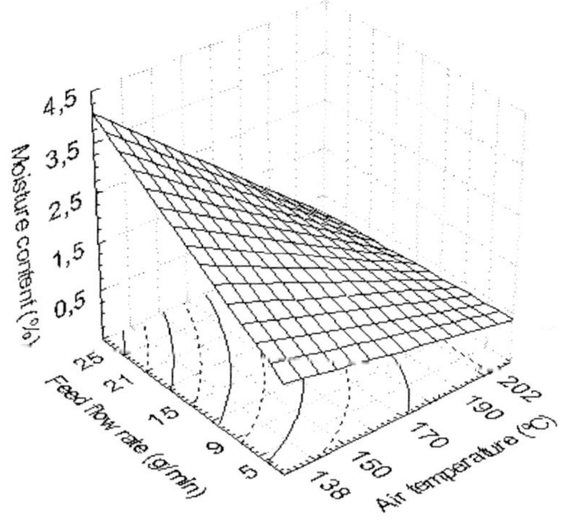

a

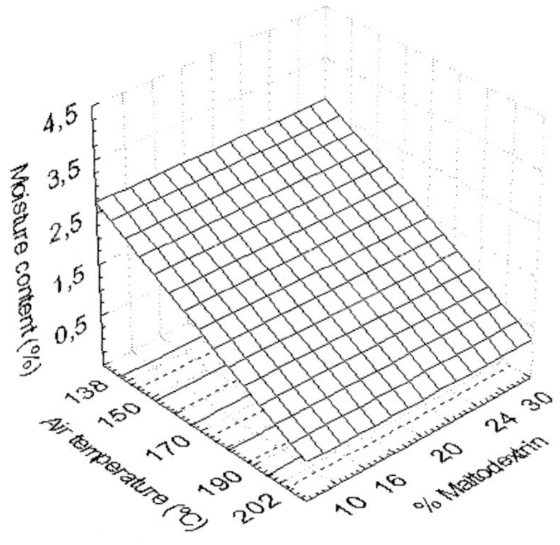

b

From Tonon et al., 2008.

Figure 2. Response surface for moisture content, for (a) 20% maltodextrin, (b) feed flow rate of 15g/min.

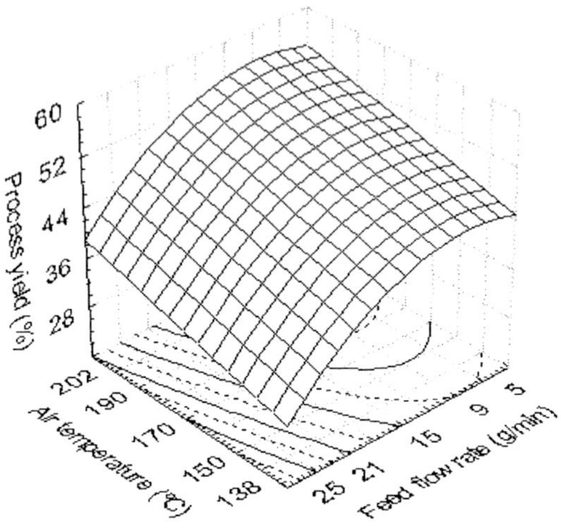

a

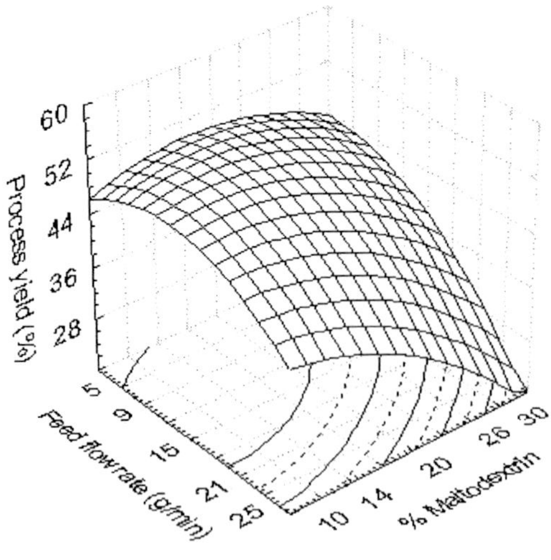

b

From Tonon et al., 2008.

Figure 3. Response surface for process yield, for (a) 20% maltodextrin, (b) inlet air temperature of 170°C.

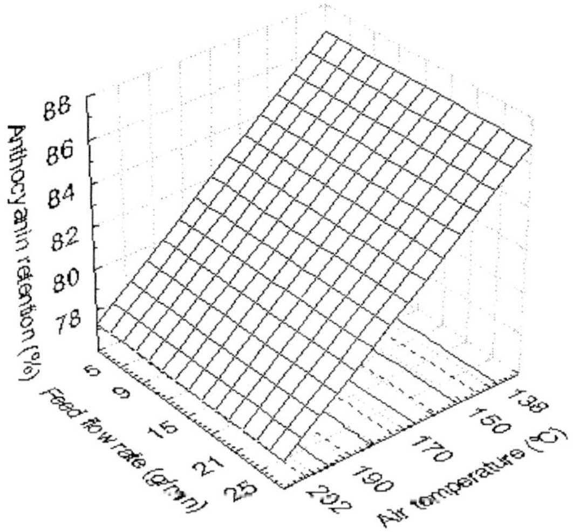

From Tonon et al., 2008.

Figure 4. Response surface for anthocyanin retention, for 20% maltodextrin.

The highest process yields (50 - 55%) were also obtained when higher drying temperatures and lower feed flow rates were used, due to the more efficient heat and mass transfer's occurrence when higher inlet air temperatures were used. As stated before, increasing inlet air temperatures and decreasing flow rates also result in powders with lower moisture content, which are less susceptible to stickiness on the dryer chamber wall. Moreover, higher temperatures lead to the fast formation of a hard crust, which makes particles less "plastic" and thus susceptible to caking and stickiness, when compared to those produced at lower temperatures, which may be more pliable and collapsed. Maltodextrin concentration also showed a negative effect on process yield, probably due to the increase on mixture viscosity, which can lead more solids to paste in the main chamber wall, thus reducing process yield (Cai & Corke, 2000).

Inlet air temperature was the only variable affecting anthocyanin retention, mainly due to the high sensitivity of these pigments to high temperatures. Moreover, powders produced at lower temperatures have a tendency to agglomerate, because of their higher moisture content. This agglomeration reduces the powder exposition to oxygen, protecting the pigments against oxidation (Quek et al., 2007).

3.2. Selection of the Best Drying Conditions

As color (as well as its associated functional properties) is the main attractive in powdered açai juice, the main criterion used for selection of the best process conditions was the anthocyanin retention. As discussed before, the only variable that influenced this response was the inlet air temperature (lower air temperatures resulted in better retention). Thus, the drying temperature chosen for further tests was 140°C.

Considering that the richest particles in anthocyanins were those produced with lower maltodextrin concentration (since there is lower "dilution effect" by maltodextrin) and that this variable did not influence pigment retention, the maltodextrin concentration selected was 10%.

Regarding to the feed flow rate, in the tests performed with feed rates higher than 15 g/min, there was a dripping inside the main chamber, that is, part of the mixture passed straight to the chamber and was not atomized, resulting in lower process yield and product wasting. On the other hand, for feed rates of 15 g/ml or lower, process yield was considered high. Thus, as lower feed rates represents higher energy spent, the feed flow rate selected as the best was of 15 g/min.

Therefore, the best process conditions selected from the central composite design were: inlet air temperature of 140°C, feed flow rate of 15 g/min and maltodextrin concentration of 10%.

3.3. Additional Tests

According to the results obtained from the central composite design, anthocyanin retention was not influenced by maltodextrin concentration. This suggests that the use of lower concentrations could result in products with higher anthocyanin content (less "diluted" by maltodextrin). Thus, some additional tests were performed with lower maltodextrin concentrations, in the selected conditions of inlet air temperature and feed flow rate (140°C and 15 g/min, respectively).

Tests were performed using 6% and 8% of maltodextrin (concentrations lower than 6% led to excessive powder stickiness on the chamber wall and insignificant process yield). Each sample was processed in triplicate and the results were statistically analyzed by the Duncan's test. They are presented in Table 5.

Table 5. Results of spray drying of açai juice using 10%, 8% and 6% maltodextrin (MD), with inlet air temperature of 140°C and feed flow rate of 15 g/min

Tests	Moisture content (%)	Process yield (%)	Anthocyanin retention (%)
10% MD	2.54 ± 0.14[a]	49.35 ± 3.52[a]	85.84 ± 2.26[a]
8% MD	2.46 ± 0.12[a]	48.60 ± 0.56[a]	84.78 ± 3.52[a]
6% MD	2.32 ± 0.05[a]	46.88 ± 1.48[a]	85.11 ± 2.48[a]

Different letters indicate significant difference between samples produced with different maltodextrin concentrations ($p \leq 0.05$).

According to Table 5, samples produced with varying maltodextrin concentrations did not show significant differences between each other with respect to moisture content, process yield and anthocyanin retention. Thus, taking into account that the powders produced with a lower amount of maltodextrin have higher anthocyanin content, the selected concentration for spray drying of açai juice was 6%.

From the conditions selected in this first part of our study (inlet air temperature of 140°C, feed flow rate of 15 g/min and 6% of carrier agent), particles were produced using different types of carrier agents (maltodextrin 10DE, maltodextrin 20DE, gum Arabic and tapioca starch) and were characterized with respect to some physical and morphological properties, physical stability at different relative humidities, as well as anthocyanin stability during storage. The results are presented in the next topics.

Chapter 4

PHYSICAL CHARACTERIZATION OF POWDERED AÇAI JUICE PRODUCED WITH DIFFERENT CARRIER AGENTS

The use of different carrier agents for powder production can result in different physicochemical properties, depending on the structure and the characteristics of each agent. According to Barbosa-Cánovas & Juliano (2005), knowledge of food properties is essential for optimizing processes and functionalities, as well as for cost reduction, especially in the case of powders produced or used in pharmaceutical and food industries.

Properties such as water activity are essential for powder stability and storage, since it represents the free water available for deterioration reactions. Density is an important property to be studied in dried mixtures. Knowledge of density is essential in industrial processes, for determination of storage, processing, packaging and distribution conditions. The products obtained by milling or drying are, in general, characterized by their bulk density, which considers the volume of the solid material and of all the porous closed or open to the atmosphere (Barbosa-Cánovas & Juliano, 2005). On the other hand, the absolute density only considers the volume of the solid material, excluding the pores' volume. Particle size distribution is also important for food powders in many aspects, such as processing, handling and shelf life, since it can influence the product taste, colour, texture and flavour (O'Hagan et al., 2005), which are the characteristics most considered by consumers. Microstructure, in turn, relates to powders' functionality, stability and flowability (Jackson & Lee, 1991; Shahidi & Han, 1993).

In this fourth chapter, we evaluated the physical properties of spray-dried açai juice produced with four different types of carrier agents: maltodextrin 10DE, maltodextrin 20DE, gum Arabic and tapioca starch. Powders were produced using the process conditions selected in the first part of our study (Chapter 3) and were characterized with respect to water activity, bulk and absolute density, intergranular porosity, particle size distribution and morphology.

Water activity was measured in an Aqualab 3TE (Decagon, Pullman, WA, USA) hygrometer, at 25°C.

Bulk density (ρ_{bulk}) was measured by weighing 2 g of sample and placing into a 50 ml graduated cylinder. Absolute density (ρ_{abs}) was determined by helium pycnometry, in an AccuPyc 1330 Automatic Gas Pycnometer (Micromeritics, Norcross, USA).

Intergranular porosity (ε) was calculated according to Equation (2):

$$\varepsilon = 1 - \frac{\rho_{bulk}}{\rho_{abs}} \quad (2)$$

Particle size distribution was measured in a laser light diffraction instrument, Mastersizer S (Malvern Instruments, Malvern, UK), using isopropanol as the dispersion medium and particle morphology was evaluated by scanning electron microscopy (SEM), performed at 5 kV, in a LEO440i scanning electron microscope (LEICA Electron Microscopy Ltd, Cambridge, UK).

Results (Table 6) were obtained in triplicate and statistically analyzed by analysis of variance, using the software STATISTICA 5.0 (StatSoft, Tulsa, USA). Mean analysis was performed using Duncan's procedure at $p \leq 0.05$.

Table 6. Water activity, bulk density, absolute density and intergranular porosity of powders produced with different carrier agents

Carrier agent	Water activity	Bulk density (g/cm^3)	Absolute density (g/cm^3)	Porosity (%)
Maltodextrin 10DE	0.229 ± 0.006^a	0.390 ± 0.015^a	1.531 ± 0.004^a	74.50 ± 1.01^a
Maltodextrin 20DE	0.245 ± 0.002^b	0.370 ± 0.016^a	1.511 ± 0.004^b	75.49 ± 1.07^a
Gum Arabic	0.244 ± 0.002^b	0.377 ± 0.011^a	1.491 ± 0.008^c	74.70 ± 0.72^a
Tapioca starch	0.189 ± 0.002^c	0.480 ± 0.005^b	1.514 ± 0.001^b	68.33 ± 0.30^b

Different letters indicate significant difference ($p \leq 0.05$).

4.1. Water Activity

All the samples showed water activity values (a_w) below 0.3, which is very positive for powder stability, once it represents less free water available for microorganism growing and biochemical reactions and hence, longer shelf life (Fennema, 1996). A similar range of a_w values were obtained by Quek et al. (2007) for spray dried watermelon powders.

The particles produced with tapioca starch showed the lowest water activity, followed by those produced with maltodextrin 10DE. Particles produced with gum Arabic and with maltodextrin 20DE showed the highest water activity values and had no significant differences between them. Such results can be attributed to the chemical structure of gum Arabic and maltodextrin 20DE, which have a high number of ramifications with hydrophilic groups, and thus can easily bind to water molecules from the ambient air during powder handling after spray drying.

4.2. Bulk Density, Absolute Density and Intergranular Porosity

The powder produced with tapioca starch exhibited the highest bulk density, whereas the others did not significantly differ between each other. This highest bulk density can be explained by the highest molecular weight of tapioca starch. Starches are composed basically by two polymers, amylose and amylopectin, the latter showing higher molecular weight. Most of the starches contain about 20-30% of amylose and 70-80% amylopectin and these values vary according to the botanical source. Tapioca starch has higher amylopectin content when compared to corn starch (83% and 72%, respectively), which explains its higher molecular weight. The heavier the material, more easily it accommodates into the spaces between the particles, thus occupying less space and resulting in a higher bulk density.

Regarding to the absolute density, all the samples exhibited similar values. According to Table 6, the particles produced with maltodextrin 10DE showed a slightly higher density, followed by those produced with maltodextrin 20DE and tapioca starch, which did not differ between each other. The sample produced with gum Arabic showed the lowest absolute density.

Absolute density corresponds to the real solid density and does not consider the spaces between particles, in contrast to the bulk density, which takes into account all these spaces. Thus, the results indicate that the samples produced with maltodextrin 10DE, maltodextrin 20DE and gum Arabic have a higher number of interparticle spaces than the sample produced with tapioca starch. These results are expressed in the calculation of intergranular porosity, which measures exactly this quantity of spaces. Porosity is an important property in the case of microcapsules where the encapsulated material is susceptible to oxidation. The larger number of spaces between particles implies in more oxygen available to degradation reactions, leading to a faster loss of the compound being protected.

4.3. Particle Size Distribution

Figure 5 shows the particle size distribution for the powders produced with the different carrier agents. Particles exhibited a large range of sizes, with diameters varying from 0.1 to 41.0 μm, and showed a bimodal distribution, i.e. a distribution with two distinct peaks, each one representing a predominant size. This is particularly interesting in the case of powders, since the "population" of smaller particles can penetrate into the spaces between the larger ones, thus occupying less space. The presence of larger particles may be attributed to an incipient agglomeration process, where the formation of irreversible link bridges leads to the production of particles of greater size.

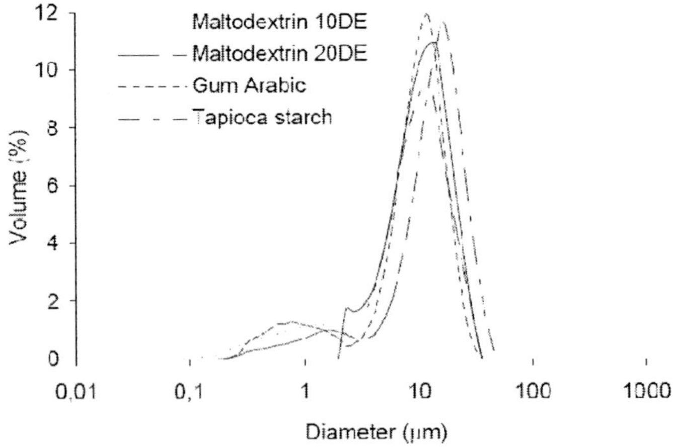

Figure 5. Particle size distribution of powders produced with different carrier agents.

The particle mean diameter varied from 9 to 14 μm, approximately, and the difference between the samples produced with the different carrier agents was small. The particles produced with gum Arabic exhibited the lowest mean diameter (9.33 μm), very similar to the particles produced with maltodextrin 20DE (9.41 μm). The sample produced with maltodextrin 10DE exhibited a mean diameter higher than both (10.94 μm), while the one produced with tapioca starch was even bigger (13.67 μm). This increased particle size is related to the molecular size of each carrier agent. The higher the maltodextrin DE, the higher is the degree of hydrolysis, and therefore the shorter are its chains. This explains the smaller particle size of the particles produced with maltodextrin 20DE, when compared to those produced with maltodextrin 10DE and with tapioca starch.

4.3. Particle Morphology

The SEM microphotographs of the powders produced with different carrier agents are shown in Figure 6. The resulting powders had particles of various sizes, for all the carrier agents, which agree with the results obtained for particle size distribution.

The particles produced with maltodextrin 10DE, maltodextrin 20DE and gum Arabic were very similar (Figures 6a, 6b and 6c), exhibiting a predominantly spherical shape, which is typical of materials produced by spray drying. Most of the particles exhibited a shriveled surface, due to the low inlet air temperature used (140°C), which leads to slower heat transfer and results in particles with more pliable and collapsed crust (Allamilla-Beltrán et al., 2005).

Most of the particles produced with tapioca starch exhibited a rounded shape and smooth surface. This type of morphology was also observed by Loksuwan (2007), when encapsulating β-carotene with native tapioca starch. Leonel (2007) evaluated the morphology of tapioca starch and observed a very similar structure to that observed here. Thus, as tapioca starch solubility is low, the rounded and smooth particles observed in Figure 6d are probably the granules of this agent which were not dissolved and did not form a "matrix" with the juice, while the shriveled ones are the particles of the dried juice.

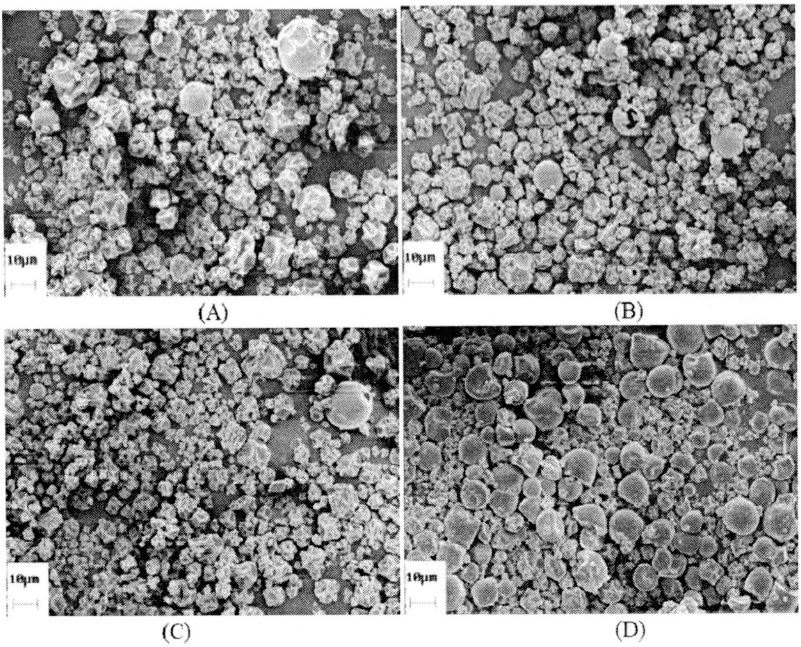

From Tonon et al., 2009b.

Figure 6. Micrographs of microcapsules produced with: (a) maltodextrin 10DE, (b) maltodextrin 20DE, (c) gum Arabic and (d) tapioca starch.

Chapter 5

STABILITY OF POWDERED AÇAI JUICE

For many years, water activity (a_w) has been considered more important than total amount of water, concerning to quality and stability of foodstuff. In this context, water sorption isotherms are important thermodynamic tools for predicting the interactions between water and food components. They describe the relationship between water activity and the equilibrium moisture content of a food product and provide useful information for food processing operations such as drying, packaging and storage (Lomauro et al., 1985).

Roos (1995) reported that the plasticization of biosolids is a result of combined effects of water and temperature. According to the author, the prediction of food stability based only on sorption isotherms data is not enough, since certain physicochemical and structural processes such as stickiness, crispness, collapse, amorphous-to-crystalline transformations and the rates of non-enzymatic browning are not related to a monolayer value and they are better correlated to the glass transition temperature through plasticization by water or temperature.

The glass transition temperature (T_g) is defined as the temperature at which an amorphous system changes from the glassy to the rubbery state. Theoretically, in the glassy state, the high viscosity of the matrix (about 10^{12} Pa · s) does not allow the occurrence of diffusion-controlled reactions (Slade & Levine, 1991). However, as the temperature increases above T_g, various changes such as increase of free volume and specific heat, as well as decrease of viscosity, are noticed (Rahman, 2006). These factors control various time-dependent structural transformations, such as stickiness, collapse and crystallization, during food processing and storage.

Thus, the use of state diagrams that indicate the material's physical state, combined with the sorption isotherms, helps in the prediction of food stability, regarding to its physical characteristics. Several authors have coupled the concepts related to water activity with those of glass transition temperature in order to evaluate food stability, thus providing an integrated approach to the role of water in food (Slade and Levine, 1991; Sablani et al., 2004; Shrestha et al., 2007; Symaladevi et al., 2009).

The critical water content/water activity is the value at which the glass transition temperature of a product is equal to the room temperature. Above the room temperature, the amorphous powders are susceptible to deteriorative changes like collapse, stickiness and caking, resulting in overall quality loss. Thus, the critical water activity (a_{wc}) indicates the maximum relative humidity to what the product can be exposed, at a given temperature, without showing structural deteriorations and diffusion-controlled reactions.

In the case of açai powder, anthocyanin degradation is one of the most important reactions affecting product's quality during storage, since these pigments are sensitive to factors like temperature, light, pH, oxygen and others. Several works are found regarding to anthocyanin stability as affected by these factors, in products like spray dried black carrot extracts (Ersus & Yurdagel, 2007), crude extracts of *Ranunculus asiaticus* flowers (Amr & Al-Tamimi, 2007), black carrot juice (Kirca et al., 2007) and others. Anthocyanin degradation usually follows first-order kinetics, i.e., anthocyanin content exponentially decreases with time.

Thus, in this fifth chapter, we evaluated the stability of spray-dried açai juice produced with the four different types of carrier agents mentioned before (maltodextrin 10DE, maltodextrin 20DE, gum Arabic and tapioca starch). From the critical storage conditions determined by coupling sorption isotherms and glass transition temperature, powders were stored at different temperatures and relative humidity for 120 days, when anthocyanin stability was evaluated.

5.1. Physical Stability

The physical stability of powders produced with different carrier agents were evaluated through the construction of sorption isotherms and determination of the glass transition temperature.

Sorption isotherms were determined by the gravimetric static method, using eight saturated salt solutions with water activies varying from 0.113 to 0.843. Experimental data were fitted to the modified BET model with three parameters (Equation 3), which is able to accurately predict the equilibrium moisture content over a larger range of water activities than the two-parameters BET model (Brunauer et al., 1938).

$$X_e = \frac{X_m C_{BET} a_w [1-(n+1)(a_w)^n + n(a_w)^{n+1}]}{(1-a_w)[1+(C_{BET}-1)a_w - C_{BET}(a_w)^{n+1}]} \tag{3}$$

where X_e is the equilibrium moisture content, X_m is the monolayer moisture content, n is the number of adsorbed layers and C_{BET} is the model constant.

The glass transition temperature was determined by differential scanning calorimetry (DSC), in a TA-MDSC-2920 (TA Instruments, New Castle, USA) equipped with a mechanical refrigeration system (RCS – refrigerated cooling accessory). Açai powder samples (about 5 mg) were placed into aluminum pans and equilibrated over saturated salt solutions in desiccators at 25°C. After equilibrium was reached, samples were hermetically sealed, weighed and taken for DSC analysis, being heated at 10°C/min from -70 to 120°C, two times (since the second scanning reduces the enthalpy relaxation of the amorphous powder that appears in the first scan). Equipment calibration was performed with indium ($T_{melting}$ = 156.6°C) and verification with azobenzol ($T_{melting}$ = 68.0°C). Dry helium, 25 ml/min, was used as the purge gas. All analyses were done in triplicate and data were treated by the software Universal Analysis 2.6 (TA Instruments, New Castle, USA). The glass transition temperature was taken as the mid point of the second-order transition that produces a step change in the heat flow at the temperature of phase transition.

5.1.1. Sorption Isotherms

The sorption isotherms fitted to the three-parameter BET model are presented in Figure 7 and they were of type III according to Brunauer's classification (Brunauer et al., 1938). This type of curve was also observed by Wang et al. (2008) for freeze-dried Chinese gooseberry and by Gabas et al. (2007) for vacuum dried pineapple added of maltodextrin and gum Arabic. The estimated parameters for the three-parameters BET model are presented in Table 7. The model showed a good fit to experimental data, with high R^2 values and satisfactory mean relative deviation modulus (E).

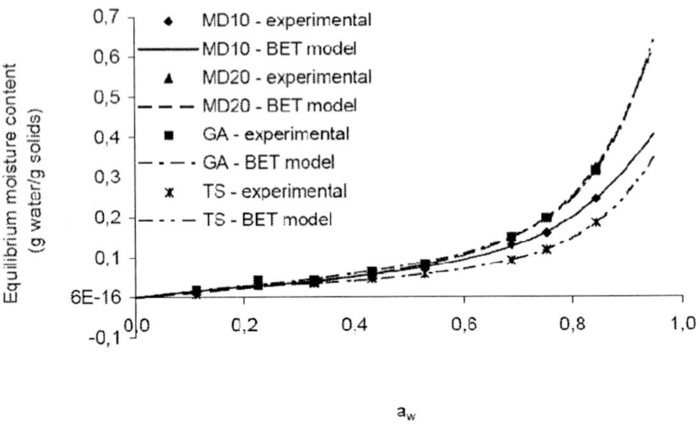

From Tonon et al., 2009a.

Figure 7. Sorption isotherms of spray dried açai juice produced with different carrier agents (MD10 = maltodextrin 10DE; MD20 = maltodextrin 20DE, GA = gum Arabic; TS = tapioca starch).

Table 7. Estimated BET parameters for açai juice powder produced with different carrier agents

Parameters	Carrier agents			
	Maltodextrin 10DE	Maltodextrin 20DE	Gum Arabic	Tapioca starch
X_m	0.045	0.058	0.054	0.031
C_{BET}	3.45	1.67	2.96	6.33
N	21.47	27.45	31.41	28.03
R^2	0.997	0.999	0.998	0.995
E (%)	6.58	8.60	6.08	15.46

The BET model is based on the monolayer moisture concept and provide the value of the monolayer moisture content of the material (X_m), considered as the safe moisture for dried foods during preservation, while most other models lack this parameter. The monolayer moisture content (X_m) indicates the amount of water that is strongly adsorbed to specific sites at the food surface and is considered an important value to assure food stability. The X_m values obtained for spray dried açai juice varied from 3.1% to 5.8% according to the BET model, which are in agreement with the values obtained by Righetto & Netto (2005) and Moraga et al. (2006), for spray-dried acerola and freeze-dried kiwi, respectively. Pérez-Alonso et al. (2006)

determined the sorption isotherms for maltodextrin 10DE and they also obtained lower monolayer moisture contents for maltodextrin 10DE (6.96-7.35%) than for gum Arabic (8.11-11.0%) at temperatures of 25, 35 and 40°C. The authors attributed such results to a combination of factors, such as the conformation and topology of molecule and the hydrophilic/hydrophobic sites adsorbed at the interface.

According to Figure 7, the powders produced with tapioca starch showed the lowest water adsorption, followed by that produced with maltodextrin 10DE, while the samples produced with maltodextrin 20DE and gum Arabic were the most hygroscopic ones. Such differences in water adsorption can be explained by the chemical structure of each agent. Maltodextrin 20DE and gum Arabic have a great number of ramifications with hydrophilic groups and therefore, can easily adsorb moisture from the ambient air. Maltodextrin 10DE is less hydrolyzed, showing less hydrophilic groups and thus adsorbing less water. Tapioca starch is a native starch (not hydrolyzed), which explains its lower hygroscopicity. Cai and Corke (2000) and Ersus and Yurdagel (2007) also verified an increase of hygroscopicity with increasing maltodextrin's DE, working with microencapsulation of betacyanins and anthocyanins, respectively. The authors attributed such increase to the lower molecular weight of the maltodextrins with higher DE, which have shorter chains and, therefore, more hydrophilic groups.

Some physical changes could be observed on the powders stored at different relative humidities. When stored at relative humidities of 43% or lower, particles remained as a free-flowing powder, for all the carrier agents used. At 53%, particles showed a beginning of agglomeration, and the powder could not flow so easily. When stored at higher relative humidities, physical transformations were more evident. At a_w's above 0.69, the particles produced with maltodextrins and gum Arabic showed the formation of hard and dark blocks, resulting from the compaction, an advanced stage in caking associated with a pronounced loss of system integrity as a result of thickening of interparticle bridges owing to flow, reduction of interparticle spaces and deformation of particle clumps under pressure (Aguilera et al., 1995). The samples produced with maltodextrin 20DE and with gum Arabic, when stored at the highest relative humidity (84%), had the appearance of a highly sticky liquid. According to Aguilera et al. (1995), in this stage of caking the interparticles bridges disappear as a result of sample liquefaction and low molecular weight fractions are solubilized. The particles produced with tapioca starch, even when stored at higher water activities, were agglomerated but did not show the formation of blocks or liquefaction, which is probably related to its lower water adsorption as compared to the powders produced with the other agents.

5.1.2. Glass Transition Temperature

The glass transition temperature decreased with increasing moisture content due to the plasticizing effect of water. The same trend was observed by Goula et al. (2008), Wang et al. (2008), Moraga et al. (2004, 2006) and Telis & Sobral (2001), working with tomato, gooseberry, kiwi, strawberry and pineapple, respectively. The plasticizing effect of water was described by the Gordon-Taylor model (Equation 4), where T_{gw} was taken as –135°C (Johari et al., 1987).

$$T_g = \frac{w_s T_{gs} + k w_w T_{gw}}{w_s + k w_w} \qquad (4)$$

where k is the constant; w is the weight fractions (g/g total); and the subscripts s and w represent solids and water.

Experimental data of T_g were well fitted to the Gordon-Taylor model, showing satisfactory values of R^2 and E. The fitted curves are shown in Figure 8. The estimated parameters are presented in Table 8.

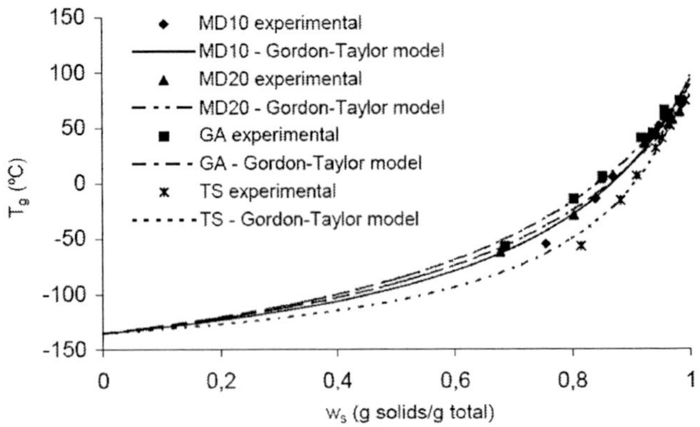

From Tonon et al., 2009a.

Figure 8. Glass transition temperature as a function of solids content for spray dried açai juice produced with different carrier agents (MD10 = maltodextrin 10DE; MD20 = maltodextrin 20DE; GA = gum Arabic; TS = tapioca starch).

Table 8. Estimated Gordon-Taylor parameters for açai juice powder produced with different carrier agents

Parameters	Carrier agents			
	Maltodextrin 10DE	Maltodextrin 20DE	Gum Arabic	Tapioca starch
T_{gs} (°C)	93.99	79.12	88.29	96.72
k	4.60	3.75	3.56	6.87
R^2	0.978	0.987	0.993	0.974
E (%)	1.83	1.49	1.32	1.89

According to Table 8, T_{gs} values varied from 79 to 97°C. The T_{gs} values obtained can explain the difficulty on drying the pure açai juice (without adding of a carrier agent). It has been shown that the addition of carrier agents such as maltodextrins and gum Arabic leads to a considerable increase on T_{gs}. Silva et al. (2006) verified an increase in the T_{gs} of freeze-dried camu-camu from 74.59 (pure pulp) to 125.45°C, when 30% of maltodextrin was added. Kurozawa et al. (2009) obtained a T_{gs} of 44.43°C for the pure chicken meat hydrolysate protein powder, while the addition of 10% of maltodextrin or gum Arabic led to T_{gs} values of 91.90 and 94.70, respectively. Thus, the spray dried pure açai juice might have a glass transition temperature much lower than the values obtained for the powders produced with additives. According to Truong et al. (2005), the sticky-point temperature is normally about 10-23°C higher than the glass transition temperature and, in spray drying, particles which are above this temperature stick to the dryer wall and degrade, and/or clump together, adversely affecting the free-flowing property. In the case of pure açai juice, considering its sugars and acids level, the sticky-temperature is much lower than 78°C (the outlet air temperature observed when the inlet air temperature was 140°C) and that would result in a high degree of stickiness and thus in an insignificant powder yield. With respect to the parameter k, the values obtained by Gordon-Taylor model were between 3.56 and 6.87, similar to those obtained for tomato, camu–camu, kiwi, garlic powder and some berries (Goula et al., 2008; Silva et al., 2006; Moraga et al., 2006; Rahman et al., 2005; Khalloufi et al., 2000). This parameter controls the degree of curvature of T_g dependence on water content (in a binary system) and can be related to the strength of the interaction between the system components (Gordon and Taylor, 1952).

The powder produced with maltodextrin 10DE showed higher glass transition temperatures as compared to that produced with maltodextrin 20DE, which is related to the decrease in the molecular weight, which decreases the T_g (Roos et al., 1996). The powder produced with tapioca starch was expected to have higher T_g than those produced with maltodextrins, since it is a native starch with higher molecular weight. However, the values obtained for such agent were similar or even lower than the obtained for the other powders. This can be attributed to the low solubility of tapioca starch at room temperature. When this agent was added to açai juice, it did not reach complete dissolution and a little quantity was precipitated inside the pipe of the spray dryer, during the process. Thus, the content of tapioca starch in the final powder was not the same that for maltodextrins and gum Arabic, which were totally soluble. As the amount of carrier agent was lower, the amount of solids content of juice was higher, which can explain the obtained T_g values, lower than the expected.

5.1.3. Critical Storage Conditions

The critical storage conditions for açai juice powder were determined by plotting the sorption isotherms and T_g data as a function of a_w, considering a room temperature of 25°C (Figure 9). The water content and T_g values were predicted by the BET and Gordon-Taylor models, respectively.

Table 9 shows the critical a_w and moisture content for the powders produced with the different carrier agents.

Table 9. Critical values for water activity (a_{wc}) and moisture content (X_c) of spray dried açai juice

Carrier agent	a_{wc}	X_c (g/g dry matter)
Maltodextrin 10DE	0.574	0.086
Maltodextrin 20DE	0.535	0.083
Gum Arabic	0.571	0.100
Tapioca Starch	0.554	0.061

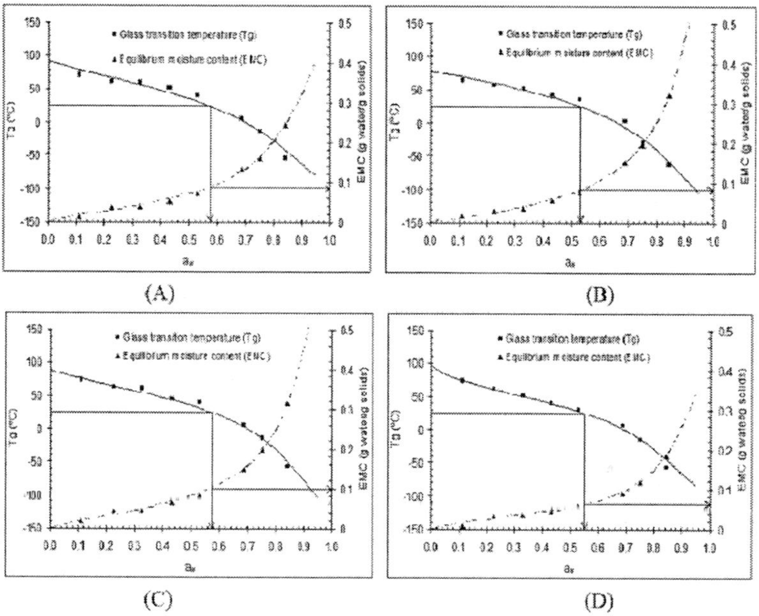

From Tonon et al., 2009a.

Figure 9. Variation of glass transition temperature (solid line) and equilibrium moisture content (dashed line) with water activity for spray dried açai juice produced with: (a) maltodextrin 10DE, (b) maltodextrin 20DE, (c) gum Arabic and (d) tapioca starch.

The critical water activities were similar for all the carrier agents, varying between 0.535 and 0.574. The powder produced with maltodextrin 10DE can be considered as the most stable, since it showed the highest a_{wc}, equal to 0.574. This means that when the powder is stored at 25°C, the maximum relative humidity to which it can be exposed is 57.4% and its moisture content is 8.6%. However, when stored at a relative humidity higher than 57.4% (at 25°C), or at a higher temperature (at a_w = 0.574), the powder will suffer physical transformations such as collapse, stickiness and caking.

Moraga et al. (2004, 2006) obtained very lower values of critical water activity and moisture content for freeze-dried kiwi (0.034% and 1.4%, respectively) and strawberry (0.110% and 2.0–4.0%) at 30°C, which is probably related to the higher sugar and acid content present in these fruits, as compared to açai. Moreover, the authors did not use any additive in the powder production, resulting in lower T_g values.

5.2. Anthocyanin Stability

For the study of anthocyanin stability upon storage, powders were put in 50 × 10 mm Petri dishes, in such a way that a large surface area was exposed during storage. The dishes were stored in airtight plastic containers filled with $MgCl_2$ and $Mg(NO_3)_2$ saturated solutions, in order to provide relative humidity values of 32.8 and 52.3%, respectively. These relative humidities were chosen based on the previously determined critical water activities of powdered açai juice. The containers were stored at two different temperatures: 25°C, representing the ambient temperature, and 35°C, which is one of the temperatures recommended by Labuza and Schmidl (1985) for accelerated shelf-life studies.

Samples were analyzed each 15 days, during 120 days, with respect to total anthocyanin, by to the spectrophotometric method. The first-order reaction rate constants (k) and half-lives ($t_{1/2}$) were calculated according to the following equations:

$$ln\left(\frac{C_t}{C_0}\right) = -kt \tag{5}$$

$$t_{1/2} = \frac{\ln 2}{k} \tag{6}$$

where C_0 is the initial anthocyanin content and C_t is the anthocyanin content at the reaction time t.

Figure 10 shows the anthocyanin degradation of powdered açai juice produced with different carrier agents, during 120 days of storage. Anthocyanin degradation exhibited two first-order kinetics: the first one, with higher reaction rate constant, up to 45-60 days of storage (here designated t_2), and the second one, after t_2, with lower degradation rate. Similar behavior was observed by Matioli & Rodriguez-Amaya (2002) and by Desobry et al. (1997), in microencapsulated lycopene and β-carotene, respectively. According to the last authors, the period with higher reaction rate corresponds to the degradation of the superficial β-carotene (non-encapsulated) or to the internal β-carotene in contact with the oxygen present in pores or entrapped in bubbles, which leads to oxidation. Moreover, at times greater than t_2, the matrix density and distance to the entrapped material would limit oxygen transfer, which explains the lower degradation rate at this period. In the same way, for açai juice powder, the higher degradation rate can be attributed to the non-encapsulated material, which shows greater contact with oxygen, or even to the material in contact with the oxygen present in the interior of pores. Moreover, the higher water adsorption at beginning of storage also can be responsible for the higher degradation rate, since higher water content implies in higher molecular mobility.

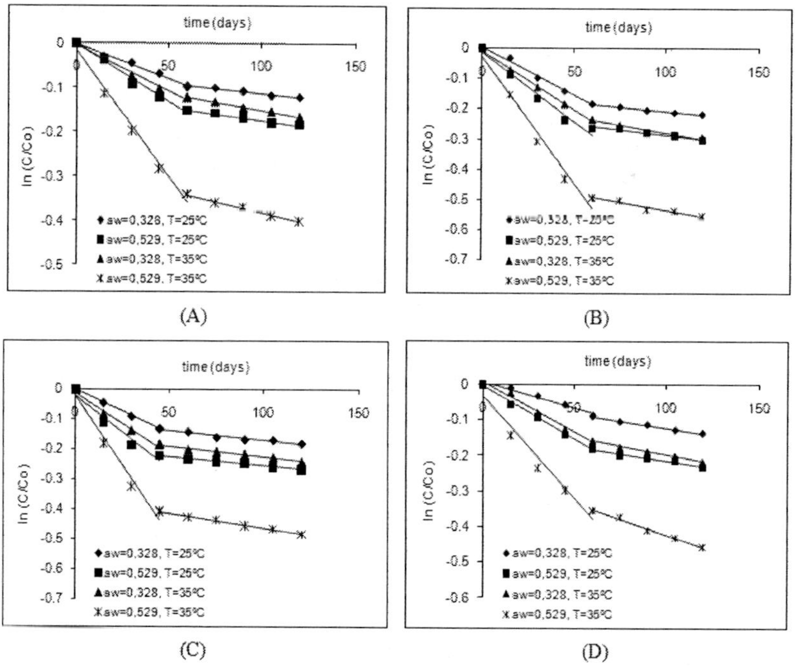

From Tonon et al., 2010.

Figure 10. Anthocyanin degradation kinetics of particles produced with: (a) maltodextrin 10DE, (b) maltodextrin 20DE, (c) gum Arabic and (d) tapioca starch.

As the particles exhibited two different first-order kinetics, two values of k and $t_{1/2}$ were calculated for each sample. However, actual half-life was determined as the time at which the anthocyanin content was reduced by 50% with respect to the zero time (Desobry et al., 1997). These values are presented in Table 10.

Table 10. Kinetic parameters of anthocyanin degradation in the powdered açai juice produced by spray drying using maltodextrin 10DE, maltodextrin 20DE, gum Arabic and tapioca starch as carrier agents

Storage conditions		Storage time	k (days^{-1})	$t_{1/2}$ (days)	R^2	Actual $t_{1/2}$ (days)
T (°C)	a_w					
Maltodextrin 10 DE						
25	0.328	$t < t_2$	0.0014	495.11	0.987	1248.29
		$t > t_2$	0.0005	1386.29	0.970	
	0.529	$t < t_2$	0.0026	266.60	0.985	962.58
		$t > t_2$	0.0006	1155.25	0.996	
35	0.328	$t < t_2$	0.0021	330.07	0.985	880.35
		$t > t_2$	0.0007	990.21	0.996	
	0.529	$t < t_2$	0.0057	121.60	0.987	411.55
		$t > t_2$	0.0010	693.15	0.987	
Maltodextrin 20 DE						
25	0.328	$t < t_2$	0.0032	216.61	0.991	909.58
		$t > t_2$	0.0006	1155.25	0.970	
	0.529	$t < t_2$	0.0045	154.03	0.971	677.64
		$t > t_2$	0.0007	990.21	0.943	
35	0.328	$t < t_2$	0.0039	177.73	0.995	516.15
		$t > t_2$	0.0011	693.15	0.946	
	0.529	$t < t_2$	0.0084	82.52	0.978	260.55
		$t > t_2$	0.0010	693.15	0.953	
Gum Arabic						
25	0.328	$t < t_2$	0.0029	239.02	0.999	978.75
		$t > t_2$	0.0006	1155.25	0.943	
	0.529	$t < t_2$	0.0049	141.46	0.950	826.25
		$t > t_2$	0.0006	1155.25	0.976	
35	0.328	$t < t_2$	0.0041	169.06	0.985	767.07
		$t > t_2$	0.0007	990.21	0.986	
	0.529	$t < t_2$	0.0091	76.17	0.976	328.55
		$t > t_2$	0.0010	693.15	0.994	
Tapioca Starch						
25	0.328	$t < t_2$	0.0015	462.10	0.963	813.81
		$t > t_2$	0.0008	866.43	0.988	
	0.529	$t < t_2$	0.0030	231.05	0.996	695.18
		$t > t_2$	0.0008	866.43	0.962	
35	0.328	$t < t_2$	0.0028	247.55	0.985	655.27
		$t > t_2$	0.0009	770.16	0.997	
	0.529	$t < t_2$	0.0057	121.60	0.959	248.14
		$t > t_2$	0.0018	385.08	0.993	

The increase of temperature led to a faster anthocyanin degradation, which was expected, since these pigments are highly thermo-sensitive.

Pacheco-Palencia et al. (2007) evaluated the anthocyanin stability in the whole, semi-clarified and clarified açai pulp, and verified a degradation rate 3.5 times higher when samples were stored at 20°C than when they were stored at 4°C. Kirca et al. (2007) also observed a strong dependence of anthocyanin degradation on storage temperature, during heating (at 70, 80 and 90°C) and storage (at 20 and 37°C) of black carrot juice with different pH's and soluble solids.

The faster anthocyanin degradation at higher temperature may also be related to the presence of sugars, together with proteins, which can result in the Maillard reaction (non-enzymatic browning), which generally occurs during food processing at high temperatures or during food storage for long time. According to Von Elbe & Schwartz (1996), the presence of sugars or products resulting from their degradation can accelerate the anthocyanin degradation, since this reaction rate follows the rate of conversion of sugars to furfural. Furfural, which is a derivative from aldopentoses, as well as hydroximetilfurfural, which is a derivative from keto-hexoses, are products resulting from the Maillard reaction that condense together with the anthocyanins, leading to the formation of compounds with brown coloration. This reaction is highly dependent on temperature, being accelerated by the presence of oxygen and occurring more frequently in fruit juices.

In a general way, the influence of temperature on the degradation rates was higher for the samples stored at the higher relative humidity, which indicates that water activity also plays an important role on anthocyanin degradation of powdered açai juice. The higher the water content, the higher the molecular mobility inside the food, which facilitates the physicochemical degradation reactions.

Garzón & Wrolstad (2001) and Cai et al. (1998) verified a considerable increase in the reaction rates and a decrease in half-lives of pelargonidin-based anthocyanins and Amaranthus betacyanin, respectively, when a_w was increased. Amr & Al-Tamimi (2007) observed a very pronounced negative effect of water activity on anthocyanin retention in crude extracts of *Ranunculus asiaticus* flowers. According to these authors, low a_w values resulted in lower conversion of anthocyanins to their hydrated carbinol base, which is less stable. This explains the greater pigment retention in the powders stored at this condition.

According to Slade & Levine, 1991, when the material is in the glassy state, the high viscosity of the matrix (about 10^{12} Pa · s) does not allow the occurrence of diffusion-controlled reactions. However, some authors have demonstrated that some diffusion-controlled reactions, such as non-enzimatic browning and lipid oxidation, may occur, even at the glassy state

(Miao & Ross, 2004; Orlien et al., 2000). This contradicts the theory of Slade and Levine (1991), according to which the storage at a temperature inferior to the glass transition temperature assures product stability. According to Schebor et al. (1999), factors like ageing of the glassy material, rotational mobility and diffusion throughout pre-existing holes (pores), due to defects and porosity in the structure, as well as the characteristic heterogeneity of food systems, can explain the occurrence of chemical reactions in foods, even in the glassy state. Thus, in the present work, although samples were stored at water activities lower than the critical aw, the diffusion of oxygen may have caused pigment degradation in the powdered açai juice particles studied.

Gradinaru et al. (2003) also verified anthocyanin degradation at temperatures below T_g, when evaluating the stability of freeze-dried anthocyanins extracted from *Hibiscus sabdariffa* flowers, stored at 40°C at different water activities. The authors pointed out to the possibility of some sort of reactant mobility in the glassy state, indicating the insufficiency of using the calorimetrically determined T_g as an absolute threshold of stability. Moreover, according to the authors, macroscopic heterogeneities in the glassy matrix, non-homogeneous distribution of water and phase separation phenomena (demixing of reactants and inert matrix) are also likely to influence the apparent reaction rates and further explain why reactions do not cease below the measured T_g.

According to Figure 10, the increase in anthocyanin degradation was much more pronounced for the powders stored at the highest temperature (35°C) and relative humidity (52.9%). As mentioned before, the water activity at which powders were exposed was chosen according to the critical a_w values observed for each sample, which were determined at 25°C. However, at a higher storage temperature, the critical a_w is lower. Thus, is it is possible that the powders stored at 35°C and 52.9% relative humidity were not in the glassy state, which can be the cause of the very higher degradation at these conditions, since molecular mobility is much greater, which can accelerate the oxidation reactions. Although there was no visual evidence that the material was in the "rubbery" state (except for samples produced with maltodextrin 20DE, which showed a beginning of agglomeration), it is important to consider that the glass transition occurs over a range of temperatures and is not defined as a specific value. Thus, it is possible that the samples stored at 35°C (at a_w of 0.529) were in this range.

Concerning to the different carrier agents used, the particles produced with maltodextrin 10DE had the highest half-life, in all the conditions studied, followed by those produced with gum Arabic. The particles produced with maltodextrin 20DE and with tapioca starch showed higher degradation rates and, consequently, lower half-lives, with respect to the others.

Rodríguez-Hernandez et al. (2005) observed better vitamin C retention in spray-dried cactus pear juice produced with maltodextrin 10DE than in the produced with maltodextrin 20DE, after drying. The authors attributed this greater retention to the better binding properties of maltodextrin 10DE, which have higher polymerization degree. Cai & Corke (2000) also observed that betacyanin retention in spray-dried *Amaranthus* pigments decreased with increasing 386 maltodextrin DE, after 16 weeks of storage. This was attributed to the higher hygroscopicity of maltodextrins with higher DE, which adsorb more water and thus are more susceptible to degradation reactions.

According to the evaluation of the powder's physical stability, the particles produced with maltodextrin 10DE had the highest critical water activity, being considered as the most stable, which was reflected in the lower anthocyanin degradation shown by them. The higher anthocyanin retention in this sample can also be related to its particle size distribution. According to Figure 5, all particles of the powder produced with maltodextrin 10DE showed diameters superior to 2 μm, while the other samples showed a little "population" of particles with mean diameter smaller than 1 μm. The smaller the particles, the larger the exposed surface area and, consequently, the faster the degradation of compounds susceptible to deterioration. In addition, the sample produced with maltodextrin 10DE had the highest absolute density, which implies in slower oxygen diffusion (Desobry, et al, 1997), thus retarding anthocyanin oxidation.

For the particles produced with gum Arabic, the second first-order kinetics started at 45 days of storage, differently from the others, in which it happened after 60 days. This can indicate a lower number of non encapsulated juice particles, or even be a consequence of the higher degradation rate observed at the first kinetics, which lead to faster degradation of the "non-protected" anthocyanins.

Although the particles produced with tapioca starch have exhibited one of the lowest half-lives, the reaction rate constant values of this sample, in the first degradation kinetics, were low and close to those of maltodextrin 10DE. This slow anthocyanin degradation can be related to the low hygroscopicity of these particles with respect to the others, which applies in lower water adsorption and thus, lower molecular mobility, making more difficult the occurrence of oxidation reactions. Moreover, the particles produced with tapioca starch had a lower porosity with respect to all the other samples, which represents lower quantity of porous containing oxygen available for pigment degradation. In the second kinetics, however, *k* values for this sample were in general higher than the obtained for the others. As previously discussed, the formation of the matrix occurs when the juice containing the dissolved carrier agent is exposed to the high temperature of the dryer chamber. Nevertheless, tapioca is a highly insoluble material and thus, it is not possible to affirm that a matrix system has been formed and that microencapsulation has occurred. Therefore, it is possible that tapioca starch has been used only as an aid for drying and that is probably why the protection provided by this agent along of storage was inferior to the other agents.

Chapter 6

CONCLUSION

Spray drying was proved to be an efficient method aiming at obtaining an anthocyanin-rich product and extending açai juice's shelf life. Inlet air temperature showed significant effect on all the responses studied. Increasing temperature led to higher process yield and to lower moisture content and anthocyanin retention. Feed flow rate negatively influenced process yield and positively influenced moisture content. The increase on maltodextrin concentration also caused a reduction on process yield, probably due to the increase on feed viscosity. The use of different types of carrier agents resulted in powders with different characteristics and different behavior during storage. The critical conditions for storage at 25°C were determined based on the sorption isotherms and the glass transition temperature. The T_g of powders decreased with increasing moisture content, confirming the strong plasticizing effect of water on this property. With respect to anthocyanin stability, pigment degradation in the spray-dried açai juice produced with different carrier agents exhibited two first-order kinetics: the first one with higher reaction rate, up to 45–60 days of storage, and the second one with lower reaction rate, after this period. Temperature negatively influenced anthocyanin stability, due to the high sensitivity of these pigments to heat, and the increase of water activity also resulted in higher degradation, due to the higher molecular mobility, which allows easier oxygen diffusion. Maltodextrin 10DE was the material that produced particles with the highest critical water activity and highest half-lives, being considered as the most adequate carrier agent for producing microencapsulated açai juice.

References

Aguilera, M.J.; Del Valle, J.M.; Karel, M. (1995). Caking phenomena in amorphous food powder. *Trends in Food Science and Technology*, 6, 149-155.

Allamilla-Beltrán, L.; Chanona-Pérez, J.J.; Jiménez-Aparicio, A.R.; Gutiérrez-López, G.F. (2005). Description of morphological changes of particles along spray drying. *Journal of Food Engineering*, 67, 179-184.

Amr, A.; Al-Tamimi, E. (2007). Stability of the crude extracts of *Ranunculus asiaticus* anthocyanins and their use as food colourants. *International Journal of Food Science and Technology*, 42, 985-991.

Barbosa-Cánovas, G.V.; Juliano, P. (2005). Physical and chemical properties of food powders. In: Onwulata, C. (Ed.) *Encapsulated and powdered foods* (1st edition, pp. 39-71). Boca Raton, USA: Taylor & Francis.

BeMiller, J.N.; Whistler, R.L. (1996). Carbohydrates. In: Fennema, O.R. (Ed.) *Food Chemistry* (3rd edition, pp. 157-223), New York, USA: Marcel Dekker.

Bhandari, B.R., Howes, T. (1999). Implication of glass transition for the drying and stability of dried foods. *Journal of Food Engineering*, 40, 71-79.

Bhandari, B.R.; Data, N.; Howes, T. (1997). Problems associated with spray drying of sugar-rich foods. *Drying Technology*, 15, 671-684.

Brunauer, S.; Emmett, P.H.; Teller, E. (1938). Adsorption of gases in multimolecular layers. *Journal of the American Chemists'Society*, 60, 309-319.

Cai, Y.Z.; Corke, H. (2000). Production and properties of spray-dried *Amaranthus* Betacyanin Pigments. *Journal of Food Science*, 65, 1248-1252.

Cai, Y.; Sun, M.; Corke, H. (1998). Colorant properties and stability of Amaranthus betacyanin pigments. *Journal of Agricultural and Food Chemistry*, 46, 4491-4495.

Cano-Chauca, M.; Stringheta, P.C.; Ramos, A.M.; Cal-Vidal, J. (2005). Effect of the carriers on the microstructure of mango powder obtained by spray drying and its functional characterization. *Innovative Food Science and Emerging Technologies*, 5, 420-428.

Coisson, J.D.; Travaglia, F.; Piana, G.; Capasso, M.; Arlorio, M. (2005). *Euterpe oleracea* juice as a functional pigment for yogurt. *Food Research International*, 38, 893-897.

Desai, K.G.H.; Park, H.J. (2005). Recent developments in microencapsulation of food ingredients. *Drying Technology*, 23, 1361-1394.

Desobry, S.A.; Netto, F.M.; Labuza, T.P. (1997). Comparison of spray-drying, drum-drying and freeze-drying for β-carotene encapsulation and preservation. *Journal of Food Science*, 62, 1158-1162.

Dib Taxi, C.M.; Menezes, H.C.; Santos, A.B.; Grosso, C.R. (2003). Study of the microencapsulation of camu-camu (Myrciaria dubia) juice. *Journal of Microencapsulation*, 20, 443-448.

Ersus, S.; Yurdagel, U. (2007). Microencapsulation of anthocyanin pigments of black carrot (*Daucuscarota* L.) by spray dryer. *Journal of Food Engineering*, 80, 805-812.

Fennema, O.R. (1996). Water and ice. In: Fennema, O.R. (Ed.) *Food Chemistry* (3rd edition, pp. 17-93), New York, USA: Marcel Dekker.

Gabas, A.L.; Telis, V.R.N.; Sobral, P.J.A.; Telis-Romero, J. (2007). Effect of maltodextrin and arabic gum in water vapor sorption thermodynamic properties of vacuum dried pineapple pulp powder. *Journal of Food Engineering*, 82, 246-252.

Garzón, G.A.; Wrolstad, R.E. (2004). The stability of pelargonidin-based anthocyanins at varying water activity. *Food Chemistry*, 75, 185-196.

Gharsallaoui, A.; Roudaut, G.; Chambin, O.; Voilley, A.; Saurel, R. (2007). Applications of spray drying in microencapsulation of food ingredients: An overview. *Food Research International*, 40, 1107-1121.

Gordon, M.; Taylor, J.S. (1952). Ideal copolymers and the second-order transitions of syntetic rubbers. I. Non-crystalline copolymers. *Journal of Applied Chemistry*, 2, 493-500.

Gouin, S. (2004). Microencapsulation: Industrial appraisal of existing technologies and trends. *Trends in Food Science and Technology*, 15, 330-347.

Goula, A.M.; Adamopoulos, K.G.; Kazakis, N.A. (2004). Influence of spray drying conditions on tomato powder properties. *Drying Technology*, 22, 1129-1151.

Goula, A.M.; Karapantsios, T.D.; Achilias, D.S.; Adamopoulos, K.G. (2008). Water sorption isotherms and glass transition temperature of spray dried tomato pulp. *Journal of Food Engineering*, 85, 73-83.

Grabowski, J.A.; Truong, V.D.; Daubert, C.R. (2006). Spray-drying of amylase hydrolyzed sweetpotato puree and physicochemical properties of powder. *Journal of Food Science*, 71, E209-E217.

Gradinaru, G.; Biliaderis, C.G.; Kallithraka, S.; Kefalas, P.; Garcia-Viguera, C. (2003). Thermal stability of *Hibiscus sabdariffa* L. anthocyanins in solution and in solid state: effects of copigmentation and glass transition. *Food Chemistry*, 83, 423-436.

Hong, J.H.; Choi, Y.H. (2007). Physico-chemical properties of protein-bound polysaccharide from *Agaricus blazei* Murill prepared by ultrafiltration and spray drying process. *International Journal of Food Science and Technology*, 42, 1-8.

Jackson, L.S.; Lee, K. (1991). Microencapsulation and the food industry. *Lebensmittel-Wissenschaft und Technologie*, 24, 289-297.

Johari, G.P.; Hallbrucker, A.; Mayer, E. (1987). The glass-liquid transition of hyperquenched water. *Nature*, 330, 552-553.

Khalloufi, S.; El-Maslouhi, Y.; Ratti, C. (2000). Mathematical model for prediction of glass transition temperature of fruit powders. *Journal of Food Science*, 65, 842-848.

Kirca, A.; Özkan, M.; Cemeroglu, B. (2007). Effects of temperature, solid content and pH on the stability of black carrot anthocyanins. *Food Chemistry*, 101, 212-218.

Kurozawa, L.E.; Park, K.J.; Hubinger, M.D. (2009). Effect of maltodextrin and gum arabic on water sorption and glass transition temperature of spray dried chicken meat hydrolysate protein. *Journal of Food Engineering*, 91, 287-296.

Leonel, M. (2007). Analysis of the shape and size of starch grains from different botanical species. *Ciência e Tecnologia de Alimentos*, 23, 579-588.

Loksuwan, J. (2007). Characteristics of microencapsulated β-carotene formed by spray drying with modified tapioca starch, native tapioca starch and maltodextrin. *Food Hydrocolloids*, 21, 928-935.

Lomauro, C.J.; Bakshi, A.S.; Labuza, T.P. (1985). Evaluation of food moisture isotherm equations. 1: Fruit, vegetable and meat products. Lebensmittel-Wissenschaft und Technologie, 18, 111-117.

Matioli, G.; Rodriguez-Amaya, D.B. (2002). Lycopene encapsulated with gum Arabic and maltodextrin: Stability study. *Brazilian Journal of Food Technology*, 5, 197-203.

Miao, S.; Roos, Y.H. (2004). Comparison of nonenzymatic browning kinetics in spray-dried and freeze-dried carbohydrate-based food model systems. *Journal of Food Science*, 69, E222-E331.

Moraga, G.; Martínez-Navarrete, N.; Chiralt, A. (2004). Water sorption isotherms and glass transition in strawberry. *Journal of Food Engineering*, 62, 315-321.

Moraga, G.; Martínez-Navarrete, N.; Chiralt, A. (2006). Water sorption isotherms and phase transitions in kiwifruit. *Journal of Food Engineering*, 72, 147-156.

O'Hagan, P.; Hasapidis, K.; Coder, A.; Helsing, H.; Pokraja, G. (2005). Particle size analysis of food powders. In: Onwulata, C. (Ed.) *Encapsulated and powdered foods* (1st edition, pp. 215-245). Boca Raton, USA: Taylor & Francis.

Orlien, V.; Andersen, A.B.; Sinkko, T.; Skibsted, L.H. (2000). Hydroperoxide formation in rapeseed oil encapsulated in a glassy food model as influenced by hydrophilic and lipophilic radicals. *Food Chemistry*, 68, 191-199.

Pacheco-Palencia, L.A.; Hawken, P.; Talcott, S.T. (2007). Phytochemical, antioxidant and pigment stability of açaí (*Euterpe oleracea* Mart.) as affected by clarification, ascorbic acid fortification and storage. *Food Research International*, 40, 620-628.

Papadakis, S.E.; Gardeli, C.; Tzia, C. (2006). Spray drying of raisin juice concentrate. *Drying Technology*, 24, 173-180.

Quek, S.Y.; Chok, N.K.; Swedlund, P. (2007). The physicochemical properties of spray-dried watermelon powder. *Chemical Engineering and Processing*, 46, 386-392.

Rahman, M.S.; Sablani, S.S.; Al-Habsi, N.; Al-Maskri, S.; Al-Belushi, R. (2005). State diagram of freeze-dried garlic powder by differential scanning calorimetry and cooling curve methods. *Journal of Food Science*, 70, E135-E141.

Rahman, M.S. (2006). State diagram of foods: Its potential use in food processing and product stability. *Trends in Food Science and Technology*, 17, 129-141.

Rattes, A.L.R.; Oliveira, W.P. (2007). Spray drying conditions and encapsulating composition effects on formation and properties of sodium diclofenac microparticles. *Powder Technology*, 171, 7-14.

Ré, M. I. (1998). Microencapsulation by spray drying. *Drying Technology*, 16, 1195-1236.

Reiniccius, G.A. (2001). Multiple-core encapsulation - The spray drying of food ingredients. In: Vilstrup, P. (Ed.). *Microencapsulation of food ingredients* (1st edition, pp.151-185). Surrey, UK: Leatherhead Publishing.

Righetto, A.M.; Netto, F.M. (2005). Effect of encapsulating materials on water sorption, glass transition and stability of juice from immature acerola. *International Journal of Food Properties*, 8, 337-346.

Rodríguez-Hernández, G.R.; González-García, R.; Grajales-Lagunes, A.; Ruiz-Cabrera, M.A. (2005). Spray-drying of cactus pear juice (*Opuntia streptacantha*): Effect on the physicochemical properties of powder and reconstituted product. *Drying Technology*, 23, 955-973.

Roos, Y.; Karel, M. (1991). Water and molecular weight effects on glass transitions on amorphous carbohydrates and carbohydrate solutions. *Journal of Food Science*, 56, 1676-1681.

Roos, Y.H. *Phase transitions in foods*. San Diego, USA: Academic Press; 1995.

Roos, Y.H.; Karel, M.; Kokini, J.L. (1996). Glass transitions in low moisture and frozen foods: Effects on shelf life and quality. Scientific status summary. *Food Technology*, 50, 95-108.

Sablani, S.S.; Kasapis, S.; Rahman, M.S.; Al-Jabri, A.; Al-Habsi, N. (2004). Sorption isotherm and the state diagram for evaluating stability criteria of abalone. *Food Research International*, 37, 915-924.

SAGRI. – Secretaria Executiva de Agricultura do Pará. *Estatística.* Available from: http://www.sagri.pa.gov.br. Acess in 26th january 2010.

Schauss, A.G.; Wu, X.; Prior, R.L.; Ou, B.; Huang, D.; Owens, J.; Agarwal, A.; Jensen, G.S.; Hart, A.N.; Shanbrom, E. (2006). Antioxidant capacity ond other bioactivities of the freeze-dried amazoniam palm berry, *Euterpe oleracea* Mart. (Acai). *Journal of Agricultural and Food Chemistry*, 54, 8604-8610.

Schebor, C.; Buera, M.P.; Karel, M.; Chirife, J. (1999). Color formation due to non-enzymatic browning in amorphous, glassy, anhydrous, model systems. *Food Chemistry*, 65, 427-432.

Shahidi, F.; Han, X.D. (1993). Encapsulation of food ingredients. *Critical Reviews in Food Science and Nutrition*, 33, 501-547.

Shrestha, A.K.; Howes, T.; Adhikari, B.P.; Bhandari, B.R. (2007). Water sorption and glass transition properties of spray dried lactose hydrolysed skim milk powder. *LWT – Food Science and Technology*, 40, 1593-1600.

Silva, M.A.; Sobral, P.J.A.; Kieckbusch, T.G. (2006). State diagrams of freeze-dried camu-camu (*Myrciaria dubia* (HBK) Mc Vaugh) pulp with and without maltodextrin addition. *Journal of Food Engineering*, 77, 426-432.

Slade, L.; Levine, H. (1991). Beyond water activity: Recent advances based on an alternative approach to the assessment of food quality and safety. *Critical Reviews in Food Science and Nutrition*, 30, 115-360.

Symaladevi, R.M.; Sablani, S.S.; Tang, J.; Powers, J.; Swanson, B.G. (2009). State diagram and water adsorption isotherm of raspberry (*Rubus idaeus*). *Journal of Food Engineering*, 91, 460-467.

Telis, V.R.N.; Sobral, P.J. (2001). Glass transitions and state diagram for freeze-dried pineapple. *Lebensmittel-Wissenschaft und Technologie*, 34, 199-205.

Tonon, R.V.; Brabet, C.; Hubinger, M.D. (2008). Influence of process conditions on the physicochemical properties of açai (*Euterpe oleraceae* Mart.) powder produced by spray drying. *Journal of Food Engineering*, 88, 411-418.

Tonon, R.V.; Baroni, A.F.; Brabet, C.; Gibert, O.; Pallet, D.; Hubinger, M.D. (2009a). Water sorption and glass transition temperature of spray dried açai (*Euterpe oleracea* Mart.) juice. *Journal of Food Engineering*, 94, 215-221.

Tonon, R.V.; Brabet, C.; Pallet, D.; Brat, P.; Hubinger, M.D. (2009b). Physicochemical and morphological characterisation of açai (Euterpe oleraceae Mart.) powder produced with different carrier agents. *International Journal of Food Science and Technology*, 44, 1950-1958.

Tonon, R.V.; Brabet, C.; Hubinger, M.D. (2010). Anthocyanin stability and antioxidant activity of spray-dried açai (*Euterpe oleracea* Mart.) juice produced with different carrier agents. *Food Research International*, doi:10.1016/j.foodres.2009.12.013.

Truong, V.; Bhandari, B.R.; Howes, T. (2005). Optimization of co-current spray drying process of sugar-rich foods. Part II – Optimization of spray drying process based on glass transition temperature. *Journal of Food Engineering*, 71, 66-72.

Von Elbe, J.H.; Schwartz, S.J. Colorants. In: Fennema, O.R. (Ed.) *Food Chemistry* (3rd edition, pp. 651-722), New York, USA: Marcel Dekker.

Wang, H.; Zhang, S.; Chen, G. (2008). Glass transition and state diagram for fresh and freeze-dried Chinese gooseberry. *Journal of Food Engineering*, 84, 307-312.

INDEX

D

E

F

G

H

I

K

L

M

N

O

P

R